Friendly Houses/Preface

In the early 1980s with the customary dearth of commissions for a young architectural practice I had designed a series of prototypical houses. These were specifically affordable houses, houses that could be built inexpensively, but without the onus attached to "cheap housing." Fundamental to the design was the notion that these dwellings should have resonance with what is deeply recognized in all peoples as "house."

If you ask a child to draw a picture of a house he will draw the same shape no matter what sort of dwelling he lives in. It does not matter if he lives in a city apartment or a mobile home the pictogram will be of a square building with a pitched roof, sometimes with a chimney and a curl of smoke ascending to a sky with a smiling sun. That is "house" and I believe that the farther the designer strays from that subconscious prototype the less it becomes house.

On that premise I designed a series of houses based on a 20 foot by 40 foot plan, a dimension we had discovered was the most practical in terms of building. This was America and all building materials there come in units of four feet, divisions or multiples of that dimension, so using that module automatically induced a cost savings. We had also discovered that dimension gave room sizes that were usable and practical. To that we added a 45° pitched roof, a diagonal cut of a piece of plywood with no waste, again a practical and cost saving gesture.

I discoverd that the possibilities within this formal envelope were nearly infinite. Like the structure of a sonnet or a fugue this order provoked innovation rather than inhibiting it. So a collection of these explorations took shape.

At that time my friends Massimo and Lella Vignelli came to visit me in my studio in Boston, Massachusetts. Always my most trusted critics and mentors I showed them the collection of designs. Massimo became very excited about them and immediately seized upon the accessibility of the design to a user. These were in his words "friendly houses" – houses that welcomed and celebrated the idea of "house" without compromising their modernity. Massimo noted that the concept, indeed, the word "friendly" as applied to an object was unique to the English language and that these houses with their simple geometry and pitched roofs embodied the idea. He christened them "Friendly Houses" and proposed a book, a portfolio of the designs and, being Massimo Vignelli world-class designer, he immediately designed the book.

As luck would have it, we actually received a commission to design a house on just the sort of restricted budget for which the "Friendly Houses" were conceived. We were presented with an existing foundation which modified the pure design of the prototypes, but the principles were still applicable. The real house became number 10 in the collection.

This is that book, a box of a portfolio of designs for Friendly Houses. That it is now being brought to realization through the good offices of Oscar Riera Ojeda is as much a tribute to the wonderful perception and talent of Massimo Vignelli as to my designs.

Peter Forbes
Firenze, February 2021

Friendly Houses/Order

Ceremony as here defined induces in objects designed to accommodate the ceremony a set of conditions which, by definition, occur again and again. The ritual drinking chalice has changed its basic form very little from early Egypt to the Wimbledon trophy or the Stanley Cup. Similarly, the symbolic attributes of a prototypical house form, if accurately observed and interpreted, establish an order which can transcend the specific house and be susceptible to universal application. An analogy can be drawn with musical forms. For example, the fugue was developed to express a particular set of musical relationships. Through the strength of the formal idea it was found to be applicable to many situations and became a musical Order. If the physical means of iconographic representation in a building type are clearly perceived they, too, can become Orders.

In the American single family house the primary Order is the roof form. Anywhere in the country a child can be asked to draw a house and, virtually without exception, he or she will draw the same shape ⌂ . Often this paradigm form is quite independent of the configuration of the dwelling in which the child actually lives, such as Navajo Indian children who live in mobile homes and New York City apartment dwellers. They draw the same shape. In the United State ⌂ is the Order of House. An Order that represents a spectrum of verities from the immediate tangible problem of getting rain off the roof to the inarticulate mythic ideal of what House should be.

The first nine houses in this opus, being based on an ideal, derive their interpretation of this order from pure geometry. The plan of each is a double 20 foot square, the end elevation is another 20 foot square surmounted by a 45 degree right triangle – the mythic "pediment." This pure prism serves as a foil against which the articulation of skylights, entries and fenestration can be played. In the 10th house the entire mythic ideal serves as a point of departure for the formal variations demanded by a specific set of real conditions.

Friendly Houses/Ceremony

Inherent in an object which derives from a mythic ideal is the importance of ceremony. Again, even as myth cannot be confused with dogma, the notion of ceremony must be detached from any static connotation of prescribed ritual. Ceremony describes the comfort and security of an oft repeated act which transcends conscious decision. Shaking hands upon meeting is such a ceremony. Within the context of House there is a universe of ceremony: entering the dwelling, greeting the visitor, gathering as a family group, gathering with non-family members for festive occasions, and a multitude of minor ceremonies specific to particular occupants.

To be a valid ideal, the prototypical house must draw its formal impetus from these ceremonies. Each important day to day act, if accurately observed, induces a vocabulary of shapes which, first of all accommodate the act on a functional level, but, furthermore, describe and explain the act on a symbolic level. For example, a front door should, for functional reasons, have protection from the weather so that the waiting visitor gains at least some degree of shelter. The form which the shelter takes, however, might reflect the ceremonial importance of entry or the primary confrontation of the private family with the public outside world or the quasi-public realm of outdoor family gathering. Similarly, circulation spaces in a house must afford reasonable access to the various areas of the dwelling, but, in addition they have a role symbolically linking or separating important acts. Hence, the ceremonial significance of moving from space to space may suggest the physical design of the circulation path itself and its representation in the total form of the building.

In the nine prototypical houses the function of entry has been accommodated by a front door, weather vestibule and overhanging porch roof. The ceremony of entry has induced a formal portal supported by cylindrical columns which acknowledge the symbolic importance of "portico" without overlaying superficial or historicist specific detail (e.g., Doric columns). The entire entrance is slightly separate from the house proper – a direct iconographic representation of the ceremony of confrontation, however friendly, of the private with public realms.

Major circulation in these houses is functionally accommodated by stairs which also, by their location and configuration, link and articulate more and less private and public spaces. The stairs are skylighted, a rational solution to the functional problem of supplying adequate light. However, it is the symbolic importance of the ceremony of moving from space to space that has determined the geometric location of the stairs. Furthermore, the introduction of a skylight has added immeasurably to the iconographic importance of the function. The external manifestation of the skylight provides a further formal representation of this important act; approaching the building, a preview of some, perhaps unknown ceremony, and, upon leaving, a vestige of past experience.

I

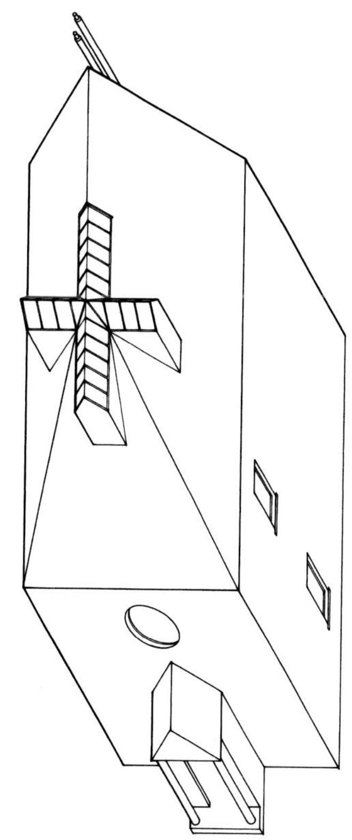

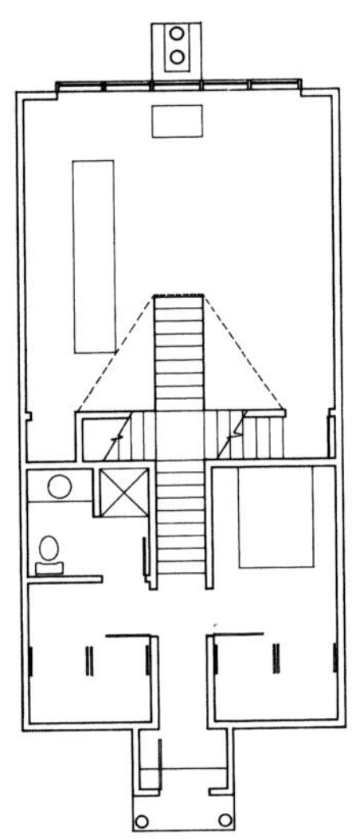

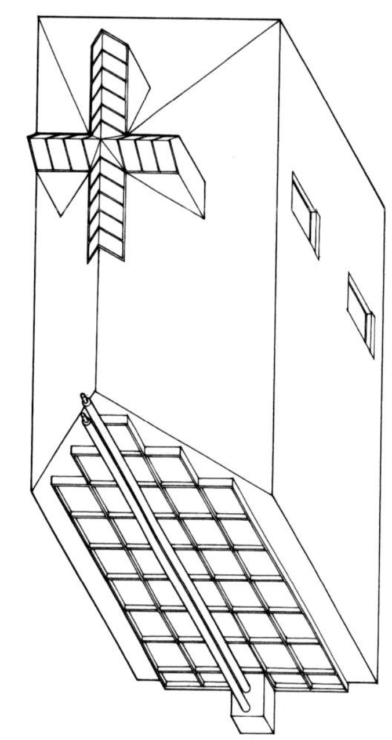

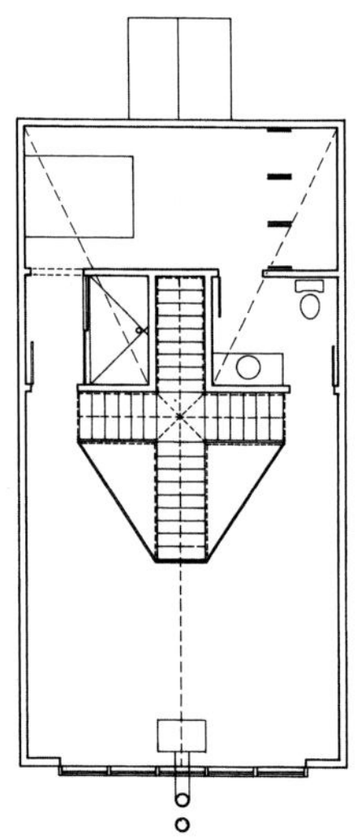

II

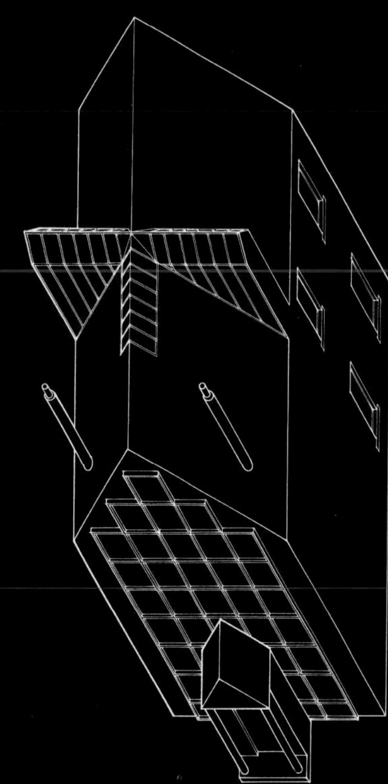

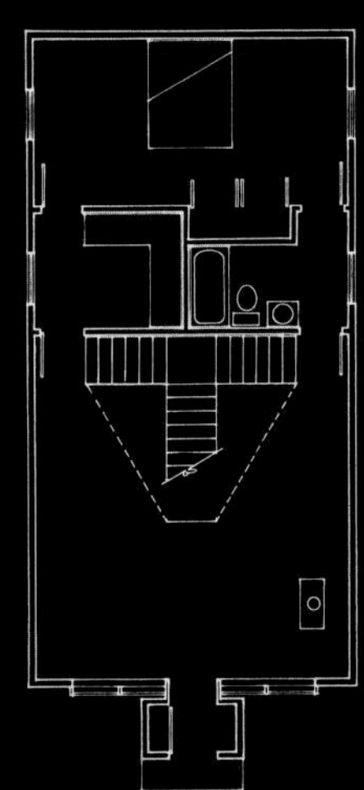

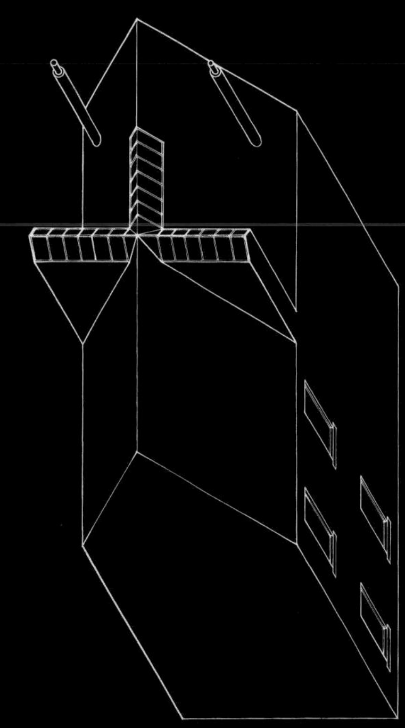

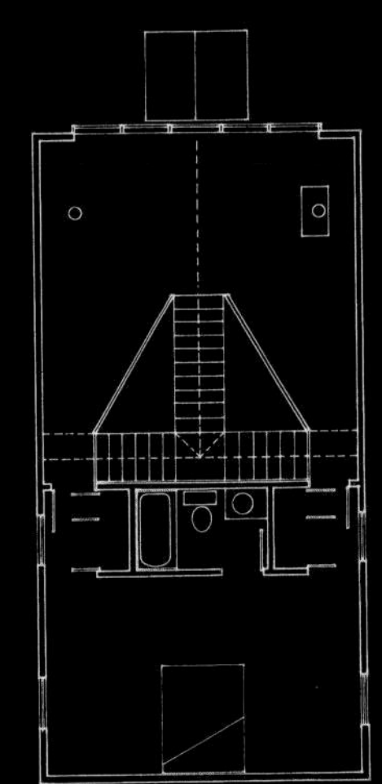

III

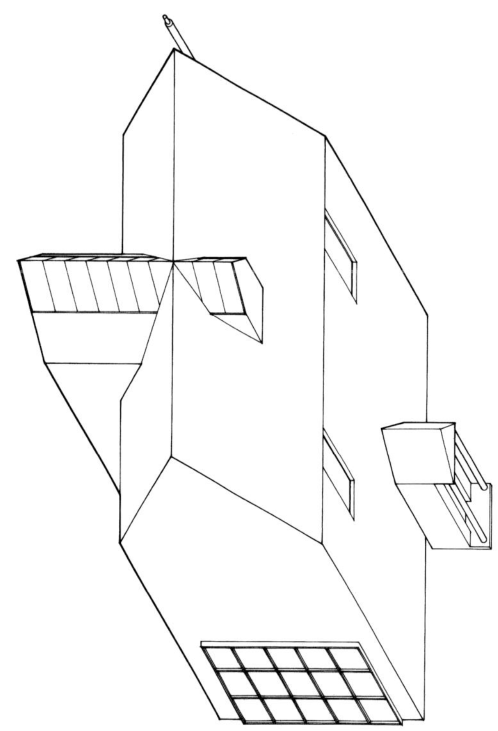

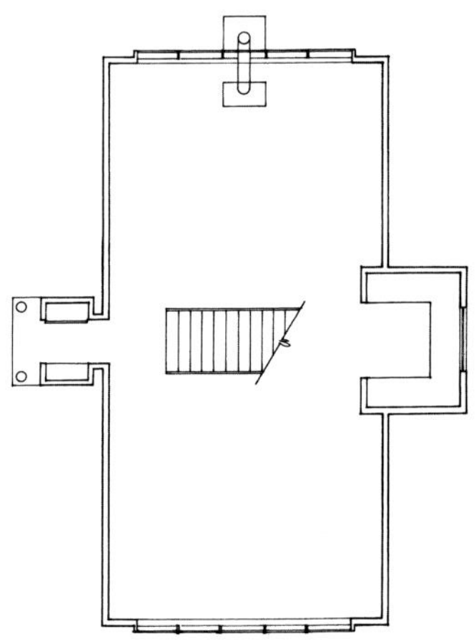

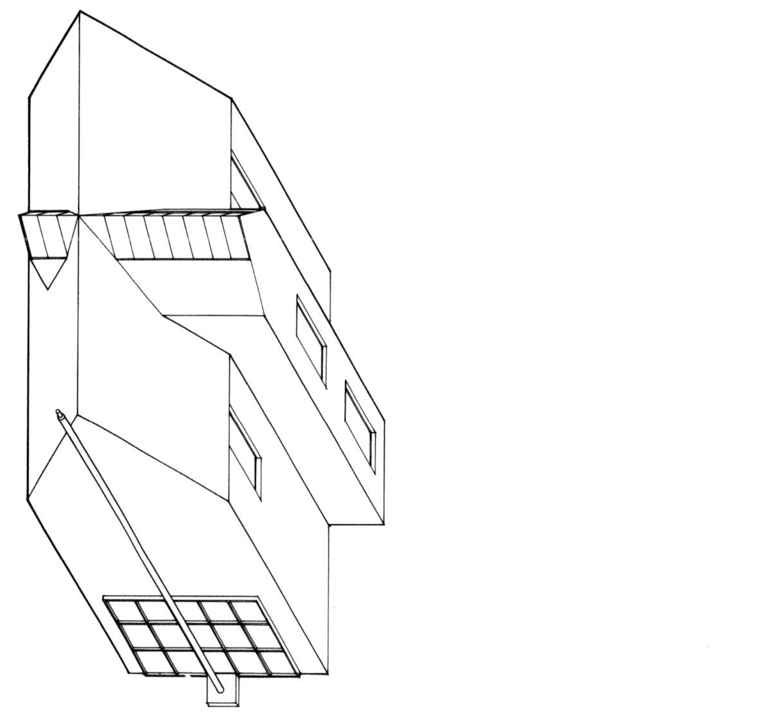

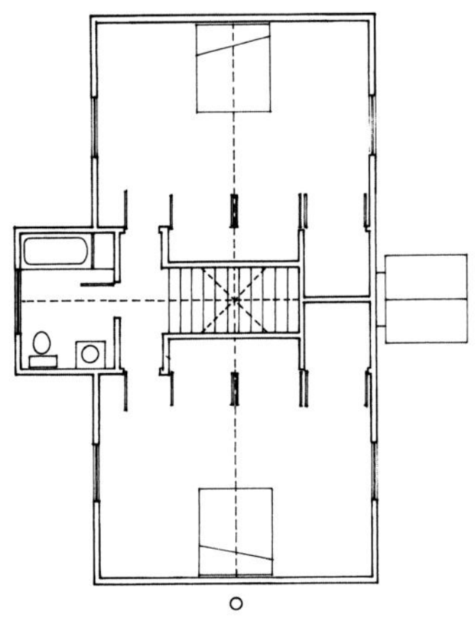

IV

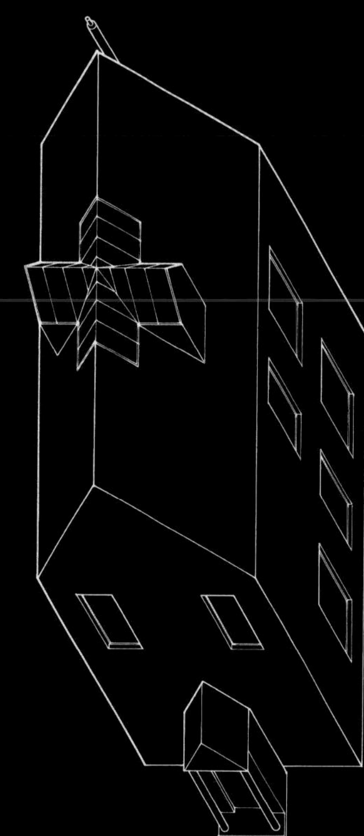

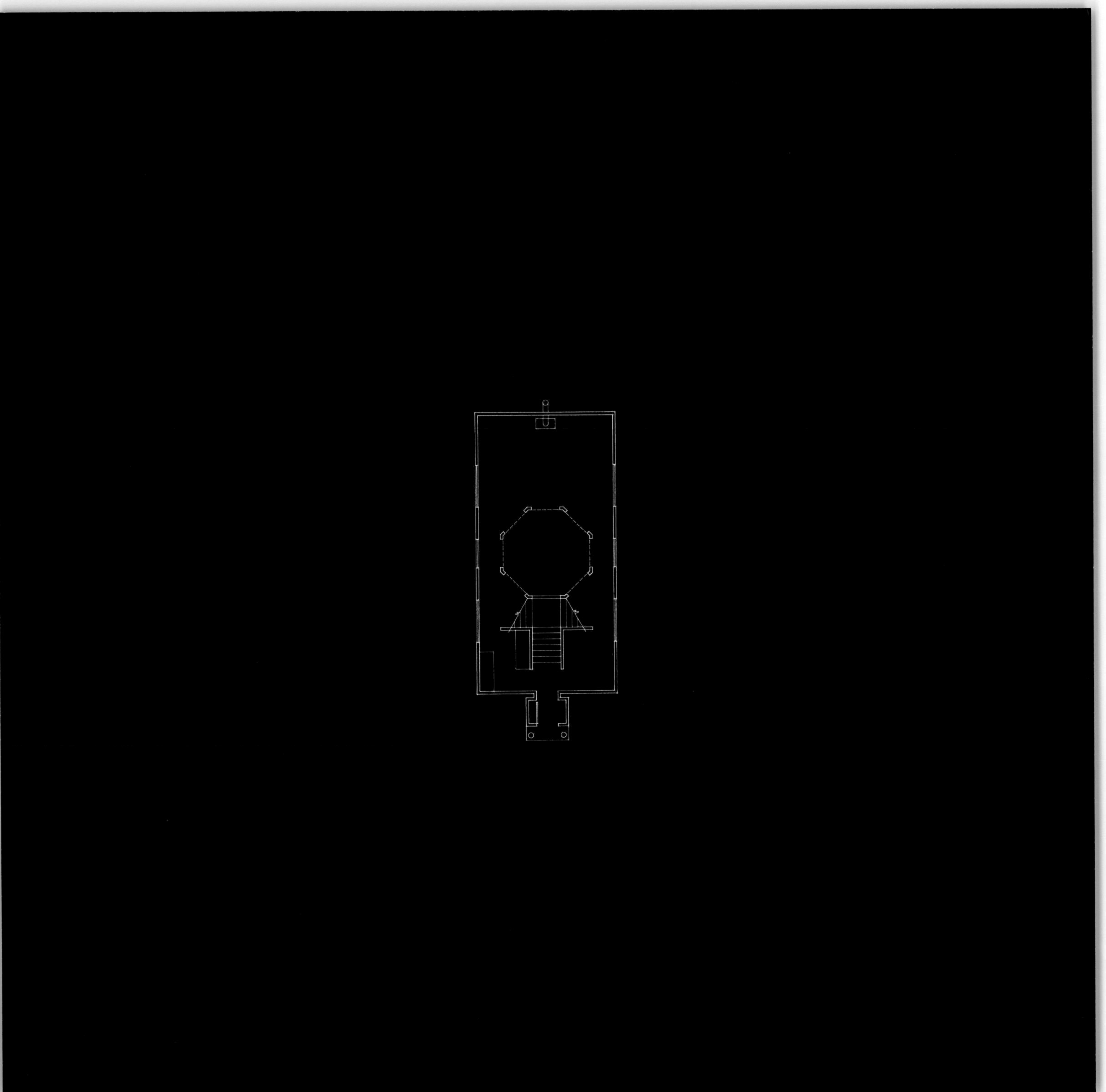

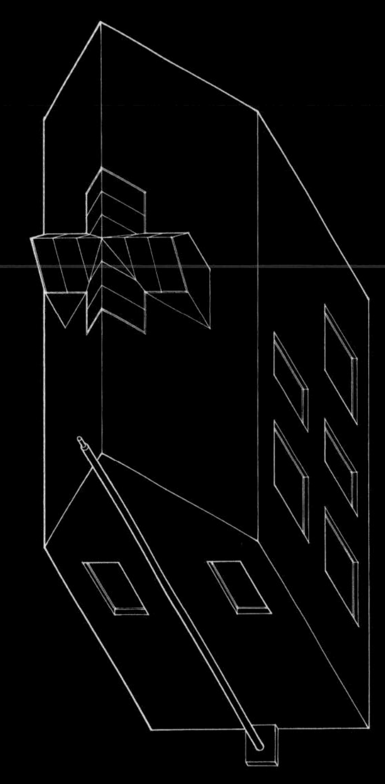

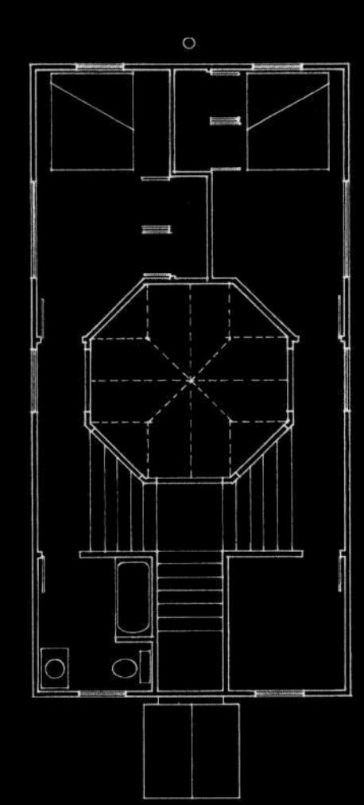

V

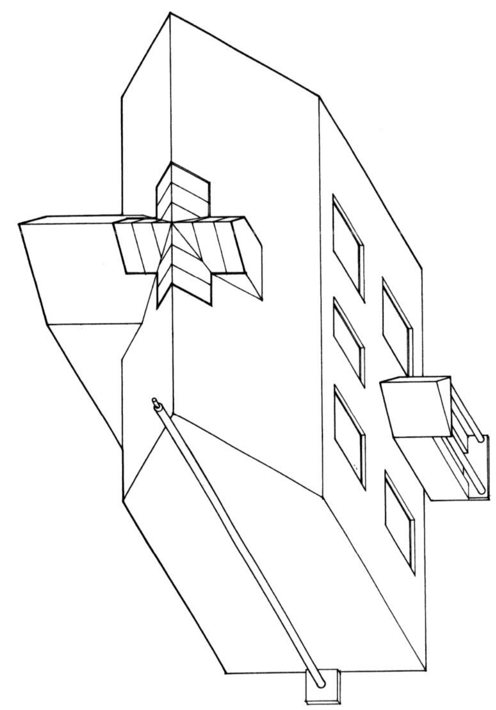

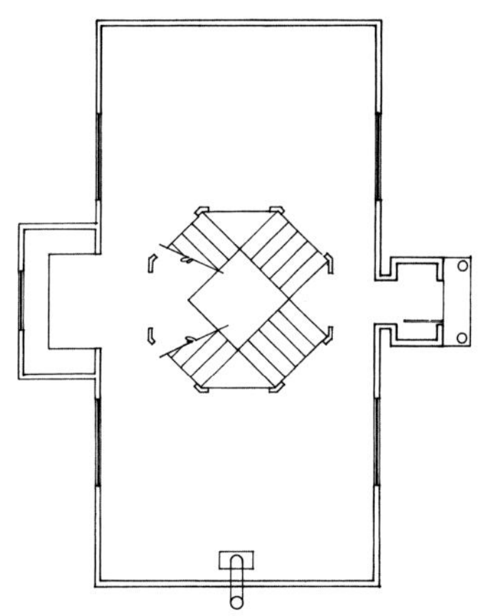

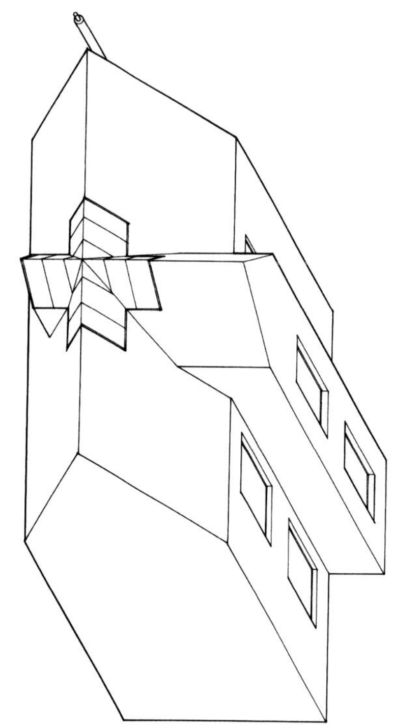

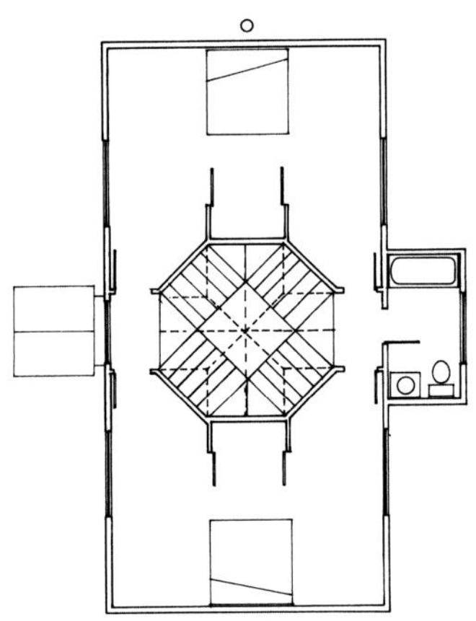

VI

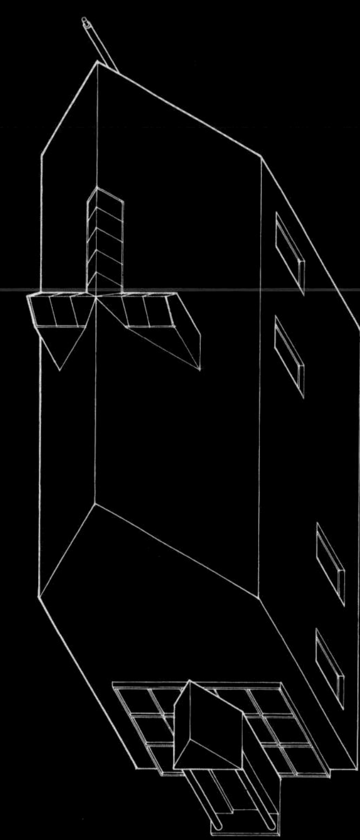

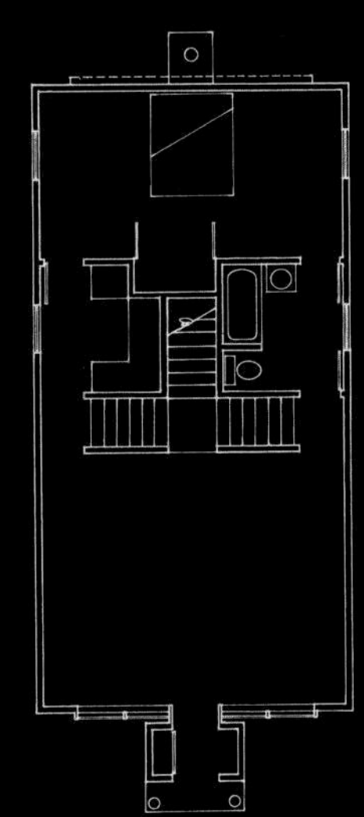

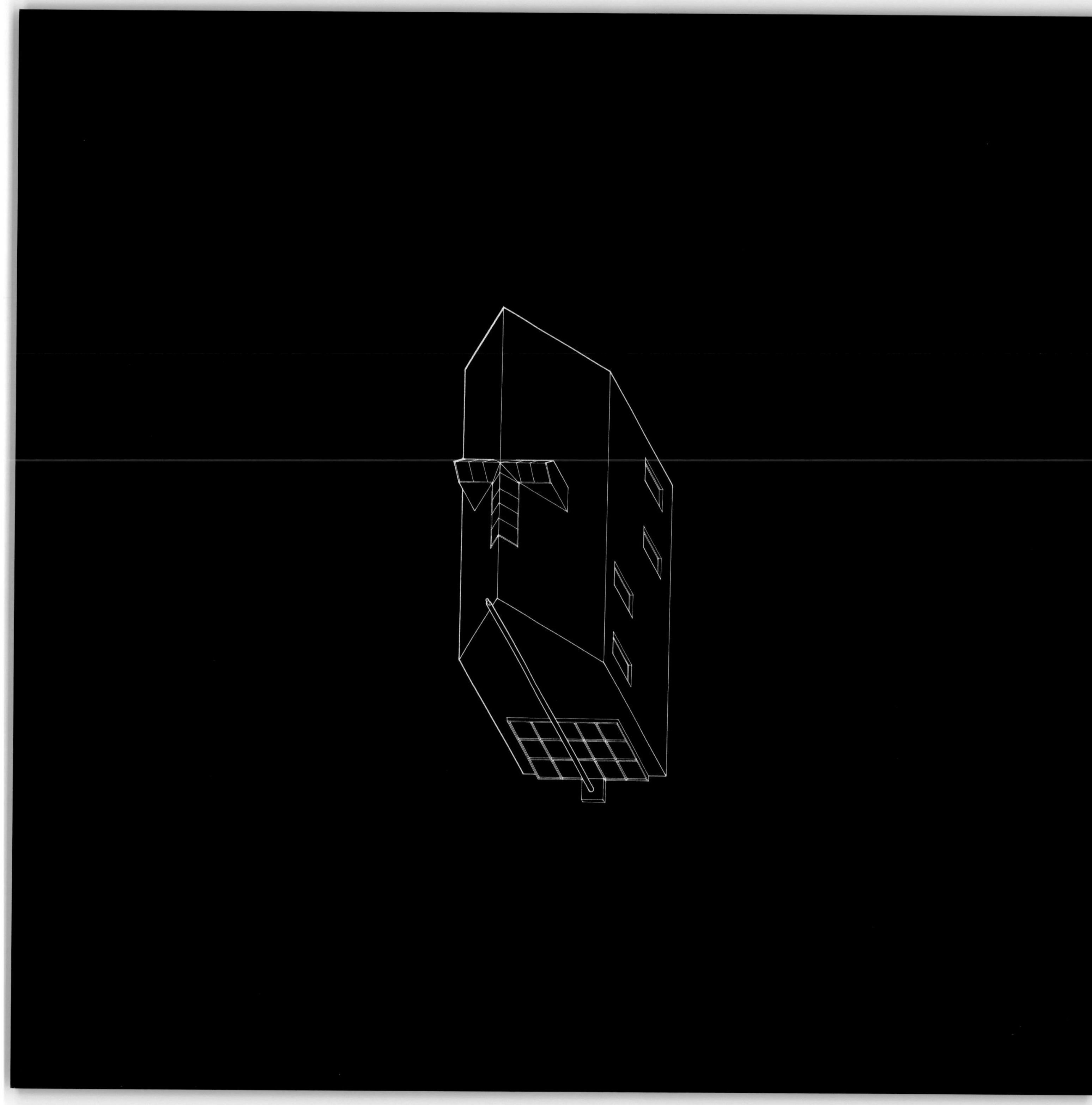

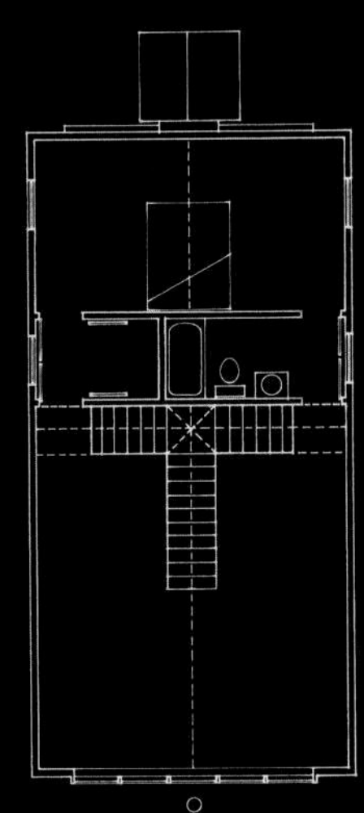

VII

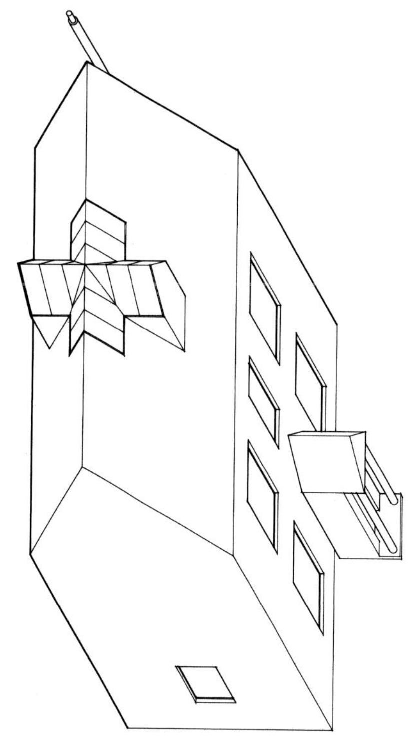

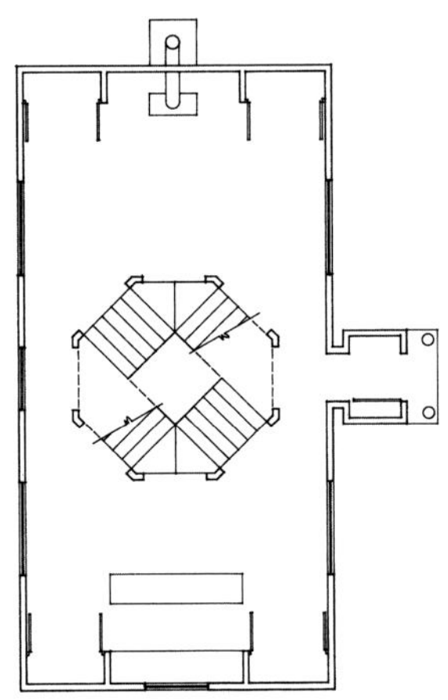

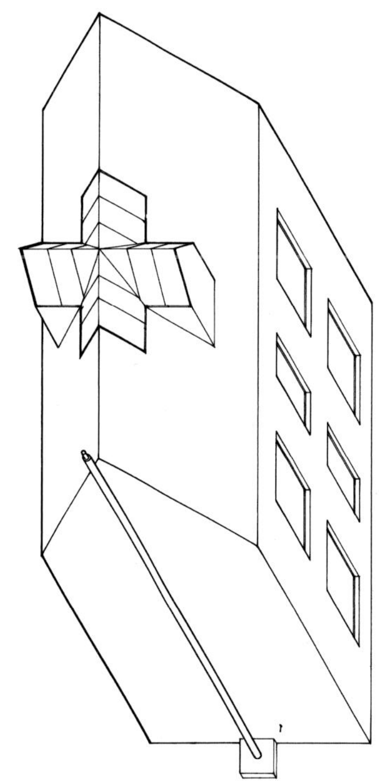

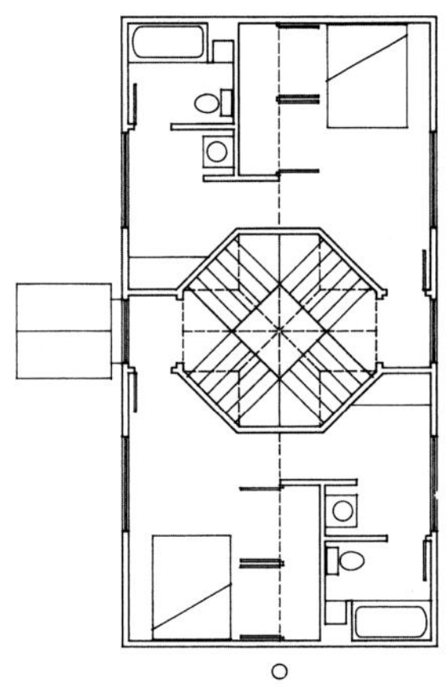

VIII

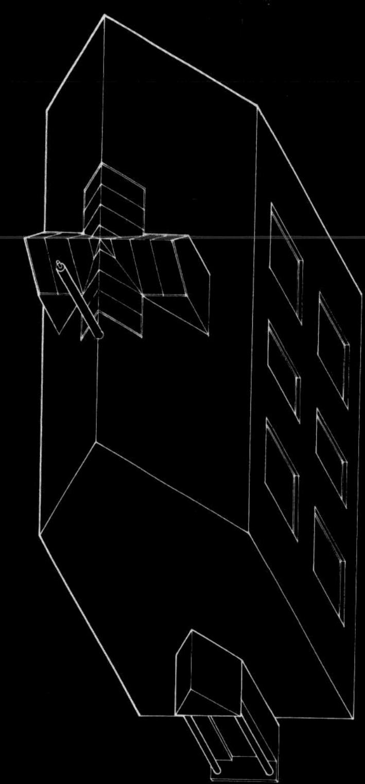

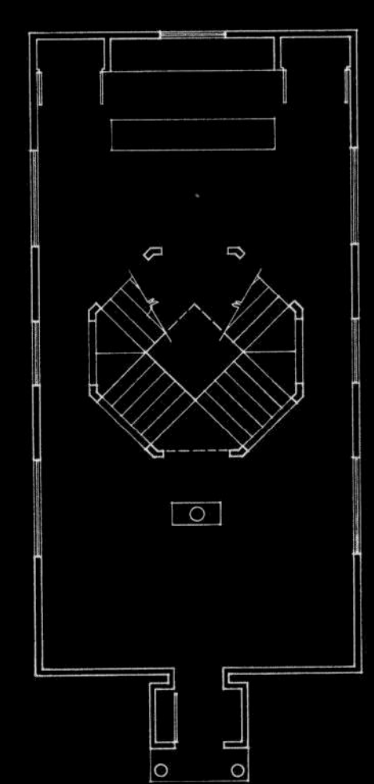

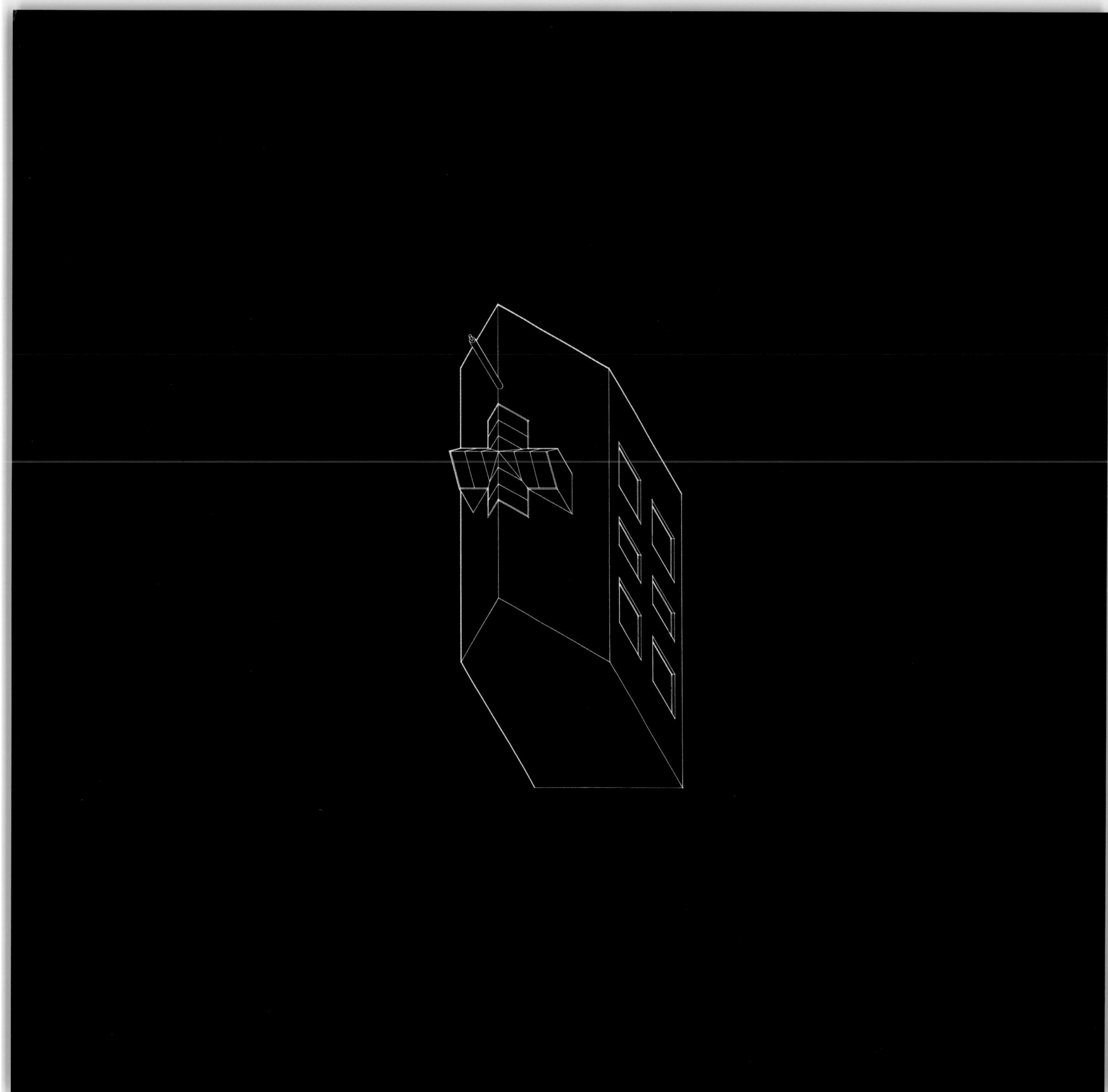

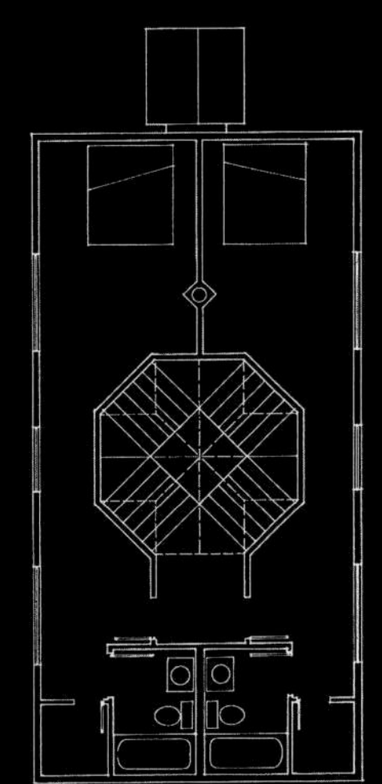

IX

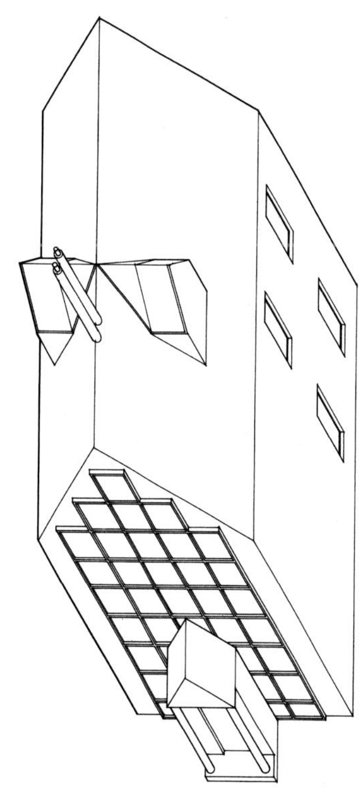

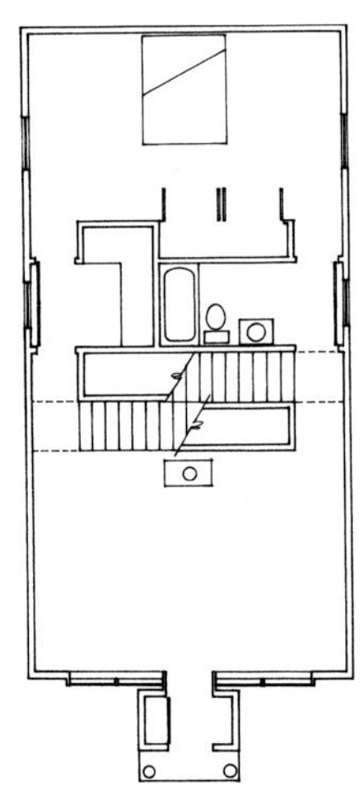

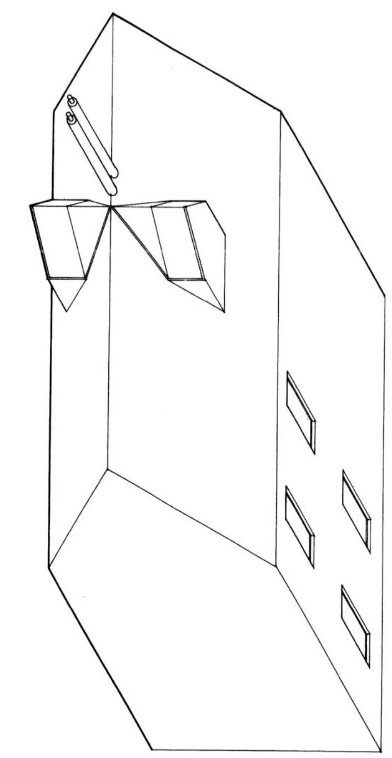

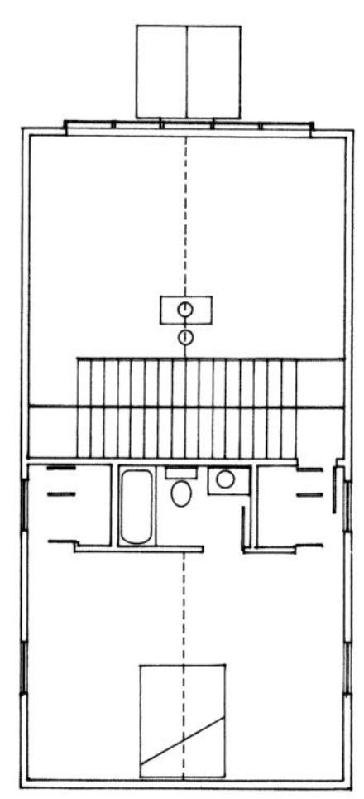

X

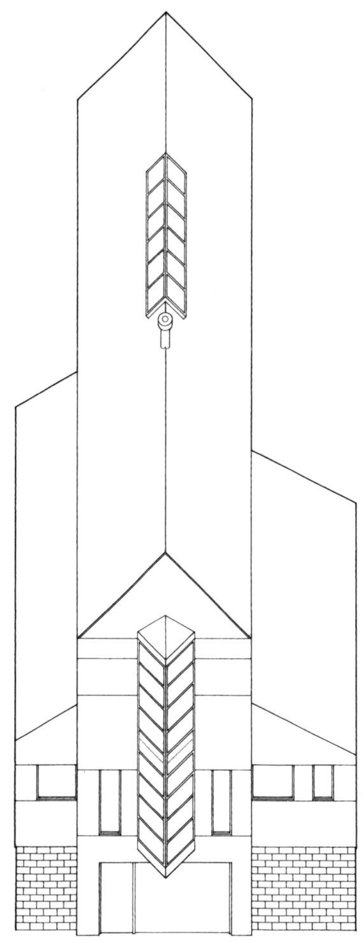

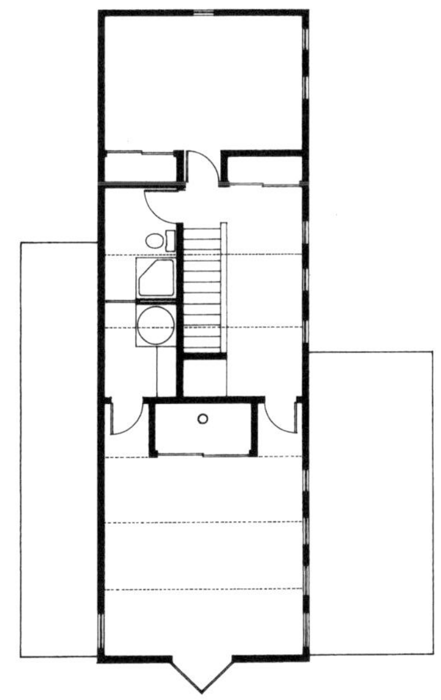

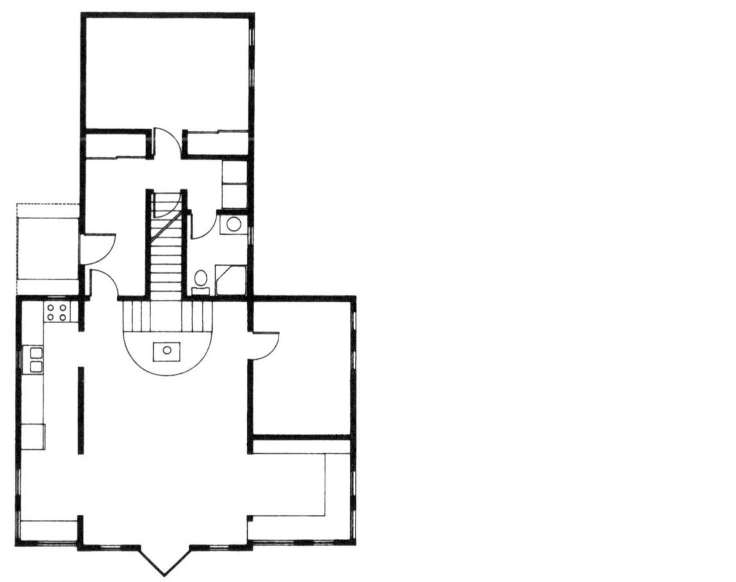

Credits

OSCAR RIERA OJEDA
PUBLISHERS

Architecture: Peter Forbes
Graphic design: Massimo Vignelli, Peter Forbes
Publication design: Oscar Riera Ojeda
Copy-editing: Kit Maude & Michael W. Phillips Jr.

ISBN 978-1-946226-49-5
Published by Oscar Riera Ojeda Publishers Limited

Oscar Riera Ojeda Publishers Limited
Unit 1003-04, 10/F.,
Shanghai Industrial Investment Building,
48-62 Hennessy Road, Wanchai, Hong Kong
T: +852-3920-9300

Production Offices | China
Suit 19, Shenyun Road,
Nanshan District, Shenzhen 518055
T:+1-484-502-5400

www.oropublishers.com | www.oscarrieraojeda.com
oscar@oscarrieraojeda.com